Beginning Statistics

+Integrated Review

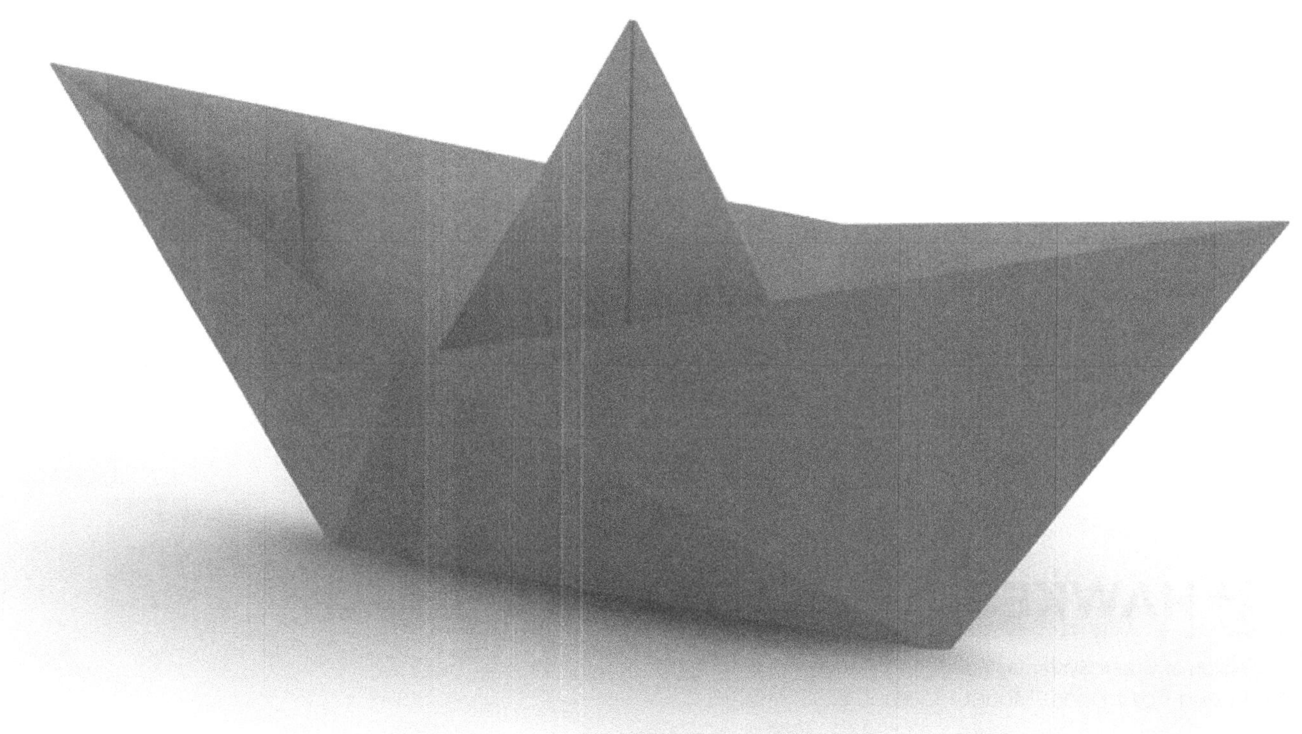

A division of Quant Systems, Inc.
546 Long Point Road, Mount Pleasant, SC 29464

ISBN: 978-1-944894-83-2

Beginning Statistics

+Integrated Review

TABLE OF CONTENTS

Chapter 1.R
Introduction to Statistics
Integrated Review

Problem Solving with Whole Numbers

○ Basic Strategy for Solving Word Problems:

1. ___________________ the problem carefully.

2. ___________________ any type of figure or ______________ that might be helpful and decide what operations are needed.

3. Perform the ________________ to solve the problem.

4. ___________________ your work.

○ Key words when solving word problems are:

Addition	Subtraction	Multiplication	Division

○ To find the average of a set of numbers, you will need to:

1. Find the __________________ of the given set of numbers.

2. ___________________ this sum by the total number of numbers in the set. This quotient is called the ________________ of the given set of numbers.

Introduction to Decimal Numbers

- Decimal notation uses a ______________________ ______________________ system and a

  ______________________ point, with whole numbers written to the

  ______________________ and fractions written to the ______________________ of the

  decimal point.

- To read or write a decimal number:

    - ______________________ (or write) the whole number.

    - Read (or write) the word "______________________" in place of the decimal point.

    - Read (or write) the ______________________ part as a whole number. Then name the fraction

      with the name of the last ______________________ on the right. Add "th" to the end of the

      fraction place value.

    - Remember that if there is not a whole number, you can put a ______________________ to the left

      of the decimal point.

- Write the following mixed number as a decimal number:

    $$3\frac{4}{100} = \underline{\quad}.\underline{\quad}\ \underline{\quad}$$

      and it would be read as ______ AND ______ ______________________

- When writing seventeen and 5 thousandths in decimal notation you would have

      ______________________ holders in the tenths and hundredths place values.

  It would look like 17.005.

- When comparing decimals:

    - Moving ______________________ to ______________________, compare digits with the same

      ______________________ value.

    - When one compared digit is larger, the ______________________ is larger.

o   Compare the following values:

5.789   Notice that the numbers are lined up, for easier comparison.
5.754

When moving from left to right, the digits are the same until the _____________________ place

values. Those are going to be used to compare. Since the 8 is a larger value than 5,

_____________________ is a larger value.

o   Rules for rounding decimals:

　　o   Look at the digit to the _____________________ of the place of desired accuracy.

　　o   If this digit is 5 or _____________________, make the digit in the desired place of accuracy one

　　larger and replace all digits to the right with zeros.

　　All digits to the left remain unchanged unless a 9 is made one larger. This effectively changes

　　the 9 to 10 which means the next digit to the left must be increased by 1.

　　o   If this digit is _____________________ than 5, leave the digit in the desired place of accuracy as

　　it is and replace all digits to the right with zeros.

　　All digits to the left remain unchanged.

　　o   Zeros **to the right of the place of accuracy** and to the right of the decimal point must be

　　_____________________. In this way the place of accuracy is clearly understood. If a rounded

　　number has a 0 in the desired place of accuracy, then that 0 remains.

Round 13.2687 to the nearest hundredth.

The digit in the hundredths is _____________________ .

The digit in the place value to the right of the hundredths is _____________________ .

Since that digit is greater than 5, the 6 in the hundredths place value changes to a 7.

The rounded value is _____._____ _____

Chapter 2.R

Graphical Descriptions of Data

Integrated Review

Introduction to Fractions and Mixed Numbers

o The parts of a fraction are (shown below):

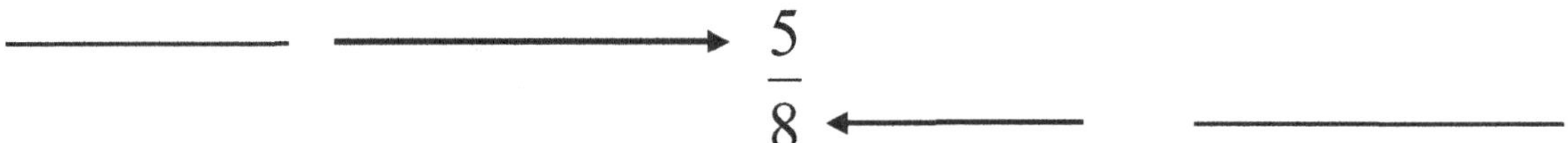

$$\frac{5}{8}$$

o Fractions are used to indicate ___________________ of a whole.

o Use the picture below to write a fraction representing the shaded portion of the shape:

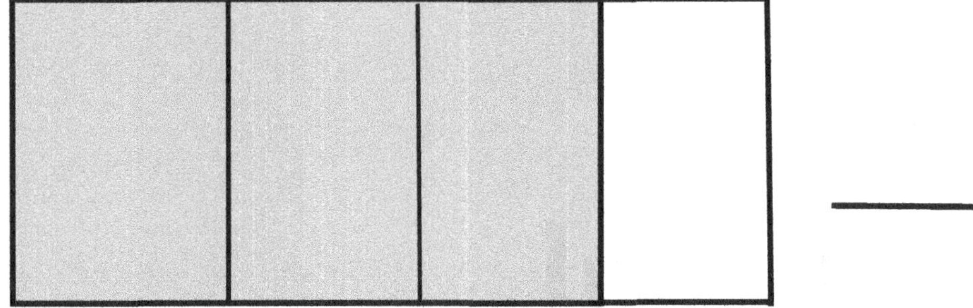

o The fraction $\frac{2}{7}$ represents _____ of _____ equal parts.

o Whole numbers can be thought of as fractions with a denominator of

___________________.

o Fraction notation represents the operation of ___________________.

o There are two rules to keep in mind when working with the value of zero in a fraction:

o For any nonzero value, b, $\frac{0}{b} = 0$. An example of this would be $\frac{0}{4} = 0$.

o For any value of a, $\frac{a}{0} =$ undefined. An example of this would be $\frac{6}{0} =$ undefined.

o Steps to multiply fractions:

    1. Multiply the ___________________.

    2. Multiply the ___________________.

- o Multiply the following fraction: $\dfrac{2}{3} \cdot \dfrac{1}{5} =$ _________________ $=$ _________

- o To find an equivalent fraction, you need to _____________________ the numerator and the

  denominator by the ____________________ nonzero whole number.

- o Find an equivalent fraction for $\dfrac{2}{3} =$ _________ by multiplying both the numerator and the denominator

  by 5.

- o There are two steps to reducing fractions to lowest terms:

  1. Factor the ____________________ and ____________________ into prime factors.

  2. Use the fact that $\dfrac{k}{k} = 1$ and divide out all of the ____________________ factors.

- o What is the common factor that can be divided out of the fraction, $\dfrac{4}{12}$?

- o What is $\dfrac{4}{12}$ in lowest terms?

- o A mixed number is the sum of a ____________________ and a ____________________

  fraction.

- o To change a mixed number to an improper fraction, you need to:

  1. Multiply the whole number by the ____________________ of the proper fraction.

  2. Add the ____________________ of the proper fraction to this product.

  3. Write this ____________________ over the denominator of the fraction.

- o Change $3\dfrac{1}{2}$ to an improper fraction:

  Multiply the whole number by the denominator: _____ · _____ = _____

  - o Add the numerator to the product from above: _____ + _____ = _____

  - o Write this sum over the denominator: _____

Decimals and Fractions

- To change from decimal numbers to fractions:

    - A terminating decimal number can be written in fraction form by writing a fraction with the following:

        - a _____________________ that consists of the whole number formed by all the digits of the decimal number

        - a _____________________ that is the power of ten that names the position of the last digit on the right

- To change from a fraction to a decimal number:

    - To change a fraction to decimal form, we divide the numerator by the denominator.

        - If the remainder is eventually 0, the decimal number is said to be _____________________.

        - If the remainder is never 0, the decimal number is said to be _____________________.

- Nonterminating decimal numbers can be _____________________ or nonrepeating.

- A _____________________ repeating decimal (also called an infinite repeating decimal number) has a repeating pattern to its digits.

- Every fraction with a whole number numerator and nonzero denominator is either terminating or repeating. Such numbers are called _____________________ numbers.

- Sometimes, changing fractions to decimal form may involve _____________________ the decimal form of a number and settling for an approximate answer. To have a more accurate answer, we may need to change the number from decimal form to _____________________ form and then perform the operations.

Decimal and Percents

- The word percent comes from the Latin *per centum*, meaning per _____________________.

- Percent means ___________________, or the ratio of a number to 100.

- The symbol _____________________ is called the percent sign. This sign has the same

  meaning as the ___________________ $\dfrac{1}{100}$.

- Changing fractions with denominators of 100 to percents:

- Example: $\dfrac{25}{100} = 25\%$

  The ___________________ did not change.

- Example: $\dfrac{3.8}{100} = 3.8\%$

  The ___________________ is not changed, the decimal point doesn't move if

  the ___________________ is 100.

- To change a decimal to a percent:

1. Move the decimal point two places to the ___________________.

2. Write the ___________________ sign.

   - Example: $0.56 = 56\%$

   - Example: $0.345 = 34.5\%$

   - Example: $0.02 = 2\%$

- To change percents to a decimal number:

1. Move the decimal two places to the ___________________.

2. Delete the ___________________ sign.

   - Example: $97\% = 0.97$

   - Example: $68.5\% = 0.685$

   - Example: $0.64\% = 0.0064$

Reading Graphs

- The Purposes of Four Types of Graphs
    - Bar Graphs: to emphasize _____________________ amounts.
    - Circle Graphs: to help in understanding _____________________ or parts of a _____________________. Circle graphs are also called _____________________ charts.
    - Line Graphs: to indicate _____________________ or _____________________ over a period of time.
    - Histograms: to indicate data in _____________________ (a _____________________ or interval of numbers).

- A common characteristic of all graphs is that they are intended to _____________________ information about numerical data _____________________ and _____________________.
- All graphs should:
    - Be clearly _____________________.
    - Be easy to _____________________.
    - Have appropriate _____________________.

- Keep these terms in mind when working with histograms:
    - Class: an _____________________ (or range) of numbers that contains data items.
    - Lower class limit: the _____________________ whole number that belongs to a class.
    - Upper class limit: the _____________________ whole number that belongs to a class.
    - Class boundaries: numbers that are halfway between the _____________________ limit of one class and the _____________________ limit of the next class.
    - Class width: the _____________________ between the class boundaries of a class (the width of each bar).
    - Frequency: the _____________________ of data items in a class.

Constructing Graphs from a Database

- o A bar graph may have either ___________________ or ___________________ bars.

  In both cases, the length of each bar represents the ___________________ of the data in

  a category being graphed.

- o When constructing a vertical bar graph, it is important to remember that the scale must be

  ___________________.

- o A circle graph is a circle that is marked in sectors (pie-shaped wedges) that correspond to

  ___________________ of data in each category represented.

- o The sum of all the central angles of the sectors in a circle graph will equal ___________.

The Real Number Line and Inequalities

o   A number line is a picture of different types of _____________________ and their

relationships to each other.

o   The graph of a number is the point that _____________________ to the number and the

number is called the _____________________ of the point.

o   On a horizontal number line, the point one unit to the left of 0 is

the _____________________ of 1. It is called negative 1 and is symbolized -1.

o   The negative sign $(-)$ indicates the _____________________ of a number as well as a

_____________________ number. It is also used to indicate _____________________.

o   The set of numbers consisting of the _____________________ numbers and their

_____________________ is called the set of integers: $\mathbb{Z} = \{..., -3, -2, -1, 0, 1, 2, 3, ...\}$.

o   The _____________________ numbers are also called _____________________ integers.

Their opposites are called _____________________ integers. Zero is its

_____________________ opposite and is neither positive nor negative $(0 = -0)$.

o   The opposite of a positive integer is a _____________________ integer, and the opposite of a

negative integer is a _____________________ integer.

o   A rational number is a number that can be written in the form of $\dfrac{a}{b}$, where a and b are

_____________________ and $b \neq$ _____________________. ($\neq$ is read "is not equal to").

o   A rational number is a number that can be written in _____________________ form as a

_____________________ decimal or as an infinite _____________________ decimal.

o   Irrational numbers can be written as infinite _____________________ decimal numbers.

- All rational numbers and irrational numbers are classified as _____________________

  numbers ($\mathbb{R}$) and can be written in some decimal form.

- Symbols of inequalities:

- $=$ is __________________ to

- $\neq$ is __________________ equal to

- $<$ is __________________ than

- $>$ is __________________ than

- $\leq$ is less than or __________________ to

- $\geq$ is greater than or __________________ to

Chapter 3.R

Numerical Descriptions of Data

Integrated Review

Addition with Real Numbers

o   The sum of two positive real numbers is ________________________.

o   The sum of two negative real numbers is ________________________.

o   The sum of a positive real number and a negative real number may be

________________________ or ________________________ (or ________________________)

depending on which number is ________________________ from 0.

o   To add two real numbers with like signs:

1.   Add their ________________________ ________________________

2.   Use the ________________________ sign

o   To add two real numbers with unlike signs:

1.   ________________________ their absolute values (the smaller from the larger)

2.   Use the sign of the number with the ________________________ absolute value

o   The ________________________ sign may be omitted when writing ________________________

numbers, but the ________________________ sign must always be written for

________________________ numbers.

o   If there is no sign in front of a real number, the real number is understood to be

________________________.

o   A number is said to be a ________________________ or to ________________________ an

equation if it gives a true statement when substituted for the variable.

Subtraction with Real Numbers

- o The opposite of a real number is called its ___________________

  ___________________.

- o The sum of a number and its additive inverse is ___________________.

- o Adding positive numbers "___________________ _______" or "___________________"

  the numbers in a ___________________ direction (moving ___________________ on the

  number line).

- o Adding negative numbers "___________________ _______" or

  "___________________" the numbers in a ___________________ direction (moving

  ___________________ on the number line).

- o In subtraction, you want to find the "____" two

  numbers.

- o To subtract, ___________________ the ___________________ of the number being

  subtracted.

- o To find the change in value between two numbers, take the end value and

  ___________________ the beginning value.

- o This process looks like:

  Change in value = (___________________ value) − (___________________ value)

Multiplication and Division with Real Numbers

o If a and b are positive real numbers, then:

   o The product of two positive numbers is ___________________ : $a \cdot b = +ab$.

      Example: $8(14) = 122$

   o The product of two negative numbers is ___________________ : $(-a)(-b) = +ab$.

      Example: $-4(-5) = 20$

   o The product of a positive number and a negative number is ___________________ : $a(-b) = -ab$.

      Example: $10(-3) = -30$

   o The product of 0 and any number is _______ : $a \cdot 0 = 0$ and $(-a) \cdot 0 = 0$.

      Example: $71(0) = 0$

o If a and b are positive real numbers (where $b \neq 0$), then:

   o The quotient of two positive numbers is ___________________ : $\dfrac{a}{b} = +\dfrac{a}{b}$

      Example: $\dfrac{14}{2} = 7$

   o The quotient of two negative numbers is ___________________ : $\dfrac{-a}{-b} = +\dfrac{a}{b}$

      Example: $\dfrac{-18}{-6} = 3$

   o The quotient of a positive number and a negative number is ___________________ :

      $\dfrac{-a}{b} = -\dfrac{a}{b}$ and $\dfrac{a}{-b} = -\dfrac{a}{b}$

      Example: $\dfrac{-144}{12} = -12$

      Example: $\dfrac{88}{-11} = -8$

o   Quick Reference Facts with Multiplication and Division with Real Numbers:

    o   If the numbers have the _____________________ sign, both the product and quotient will be

        _____________________.

    o   If the numbers have _____________________ signs, both the product and quotient will be

        _____________________.

Exponents and Order of Operations

- When looking at 5^3, you have the following parts:

  - The base is: _______________

  - The exponent is: _______________

  - The product is: _______________

  - The exponential expression is: _______________

- In expressions with exponent 2, the base is said to be _________________.

- In expressions with exponent 3, the base is said to be _________________.

- With other exponents, the base is said to be "_____________________________________."

- For any number, a, $a^1 = a$. An example would be $7^1 =$ _______________.

- For any nonzero number, a, $a^0 = 0$. An example would be $7^0 =$ _______________.

- The rules for order of operations are:

  1. Simplify within _______________ _______________, such as parentheses ( ), brackets [ ], or braces { }. Start with the _____________ group.

  2. Evaluate any numbers or expressions indicated by _____________.

  3. Moving from _____________ to _______________, perform any _____________ or _______________ in the order in which it appears.

  4. Moving from _____________ to _______________, perform any _____________ or _______________ in the order in which it appears.

Evaluating Algebraic Expressions

- o The expression $-x^2$ will have a _________________ value.

    - ▪ An example of this is $-6^2 =$ _____________ .

- o The expression $(-x)^2$ will have a _______________ value.

    - ▪ An example of this is $(-6)^2 =$ __________________ .

- o To evaluate an algebraic expression, you will need to:

    1.

    2.

    3.

Evaluating Radicals

o   Complete the tables of Squares and Perfect Squares below:

Integers (n)	1	2	3	4	5	6	7	8	9	10
Perfect Squares (n^2)										

Integers (n)	11	12	13	14	15	16	17	18	19	20
Perfect Squares (n^2)										

o   Finding the square root of a number is the opposite of _________________ a number.

o   The symbol _____________ is called a radical sign.

o   The number under the radical sign is called the _______________.

o   The complete expression, such as _______________ , is called a radical or radical expression.

o   If a is a nonnegative real number, then _________________ is the principal square root of a.

o   If a is a nonnegative real number, _____________ is the negative square root of a.

o   A number is cubed when it is used as a factor 3 times.

o   Complete the cubes of numbers below:

$1^3 =$	$2^3 =$	$3^3 =$	$4^3 =$	$5^3 =$
$6^3 =$	$7^3 =$	$8^3 =$	$9^3 =$	$10^3 =$

o   If a is a real number, then _______________ is the cube root of a.

o   In the cube root expression $\sqrt[3]{a}$, the number 3 is called the _____________.

o   In a square root expression such as $\sqrt{a}$, the index is understood to be 2 and is

______ written.

o   Expressions with square roots and cube roots, as well as other roots, are called

_______________________________ .

Probability, Randomness, and Uncertainty

Integrated Review

Multiplication and Division with Fractions and Mixed Numbers

o   To multiply with mixed numbers, you will need to change the mixed numbers to _____________________

_____________________ and then multiply the fractions.

o   You can multiply fractions and mixed numbers while reducing at the same time by using

_____________________ factors.

o   Multiply and reduce to lowest terms:

$$1\frac{2}{5} \cdot \frac{5}{8} = \frac{\quad \cdot \quad}{\quad \cdot \quad \cdot \quad} = \frac{\quad}{\quad}$$

o   Remember that if all of the factors in the _____________________ or the _____________________ divide

out, then the value of ______ must be used as a factor.

o   The reciprocal of $\frac{a}{b}$ is $\frac{b}{a}$ (when a and b do not equal 0).
The product of a nonzero number and its reciprocal is always ______.

o   Match the following values and its reciprocal:

15   2

$\frac{7}{8}$ $\frac{1}{15}$

$\frac{1}{2}$ $\frac{8}{7}$

o   To divide fractions by any nonzero number, you will _____________________ by its reciprocal.

For example, $\frac{1}{2} \div \frac{2}{5} = \frac{1}{2} \cdot \frac{\quad}{\quad}$.

Least Common Multiple (LCM)

- The _________________ of a number are the products of that number with the counting numbers.

- The least common multiple (LCM) of two (or more) whole numbers is the _____________ number that is a multiple of each of these numbers.

- Steps to find the LCM of Counting Numbers:

   1. Find the _________________ _________________ of each number.

   2. Identify the prime factors that appear in _______________ one of the prime factorizations.

   3. Form the product of these primes using each prime the _________________ number of times it appears in any one of the prime factorizations.

- Find the LCM of 24 and 36.

   1. The prime factorization for 24 is: _______________ .

      The prime factorization for 36 is: _______________ .

   2. You will use the factors of _______________.

   3. Multiply the factors together to get the LCM of _________________ .

Addition and Subtraction with Fractions

- To **add** fractions with the **same denominator**, you will need to:

  - __________________ the numerators.

  - Keep the __________________ denominator.

  - __________________, if possible.

- To **add** fractions with **different denominators**, you will need to:

  - Find the least common __________________ (LCD). Remember that the LCD is the least common multiple for the denominators.

  - Change each fraction into an __________________ fraction with that denominator.

  - __________________ the new fractions.

  - Reduce, if possible.

- A common error that occurs when adding fractions that needs to be avoided is __________________ out across the addition sign.

- To **subtract** fractions with the **same denominator**, you will need to:

  - __________________ the numerators.

  - Keep the __________________ denominator.

  - Reduce, if possible.

- To **subtract** fractions with **different denominators**, you will need to:

  - Find the least common __________________ (LCD).

  - Change each fraction into an __________________ fraction with that denominator.

  - __________________ the new fractions.

  - Reduce, if possible.

Fractions and Percents

- If a fraction has denominator _________________, it can be changed to a percent by writing the _________________ and adding the _________________ sign.

- If the denominator is a factor of _________________ (2, 4, 5, 10, 20, 25, or 50), the fraction can be changed to an equivalent fraction with denominator of _________________ and then changed to a percent.

- When fractions do NOT have factors of 100 as denominators you will need to:

- Change the fraction to decimal form by _________________, either by long division or with a calculator (depending on instructions from instructor).

- Move the decimal two places to the _________________ and write the _________________ symbol.

$$\text{Example: } \frac{3}{4} = 4\overline{)3} = 0.75 = 75\%$$

- Helpful calculation tips:

    - The numerator goes _________________ the division symbol and the denominator goes _________________ of the division symbol.

    - If you are using a calculator, you will type the numerator _________________, then the division symbol, and then the _________________, followed by enter/equal.

        - To change a percent to a fraction or a mixed number:

    - Write the percent as a fraction with 100 as the _________________ and drop the _________________ symbol.

    - _________________ the fraction, if possible.

$$\text{Example: } 80\% = \frac{80}{100} = \frac{2 \cdot 2 \cdot 2 \cdot 2 \cdot 5}{2 \cdot 2 \cdot 5 \cdot 5} = \frac{4}{5}$$

Chapter 5.R
Discrete Probability Distributions
Integrated Review

Order of Operations with Real Numbers

○   Order of Operations:

    ○   Simplify within grouping symbols, such as

       _____________________ ( ), _____________________ [ ], and

       _____________________ { }, working from the innermost grouping

       _____________________.

    ○   Find any powers indicated by _____________________.

    ○   Moving from _____________________ to _____________________, perform any

       _____________________ or _____________________ in the order they appear.

    ○   Moving from _____________________ to _____________________, perform any

       _____________________ or _____________________ in the order they appear.

○   Keep in mind that there are other _____________________ symbols like the

   _____________________ _____________________ bars (such as $|3+5|$), the

   _____________________ _____________________ (as in $\dfrac{14+7}{3}$), and the

   _____________________ _____________________ symbol (such as $\sqrt{40+9}$).

○   Even though the mnemonic PEMDAS is helpful, remember that multiplication

   and division are performed as they _____________________, left to right.

○   Also, addition and subtraction are performed as they _____________________, left to right.

Solving Linear Inequalities

- A ________________ is a collection of objects or numbers.

- The items in the set are called ___________________.

- Sets are indicated with ________________ , that look like ___________________.

- Elements are listed within the braces the set is said to be in ____________ form.

- The symbol $\in$ is read "is an ____________ of" and is used to indicate that a particular number belongs to a ______________.

- If the elements in a set can be __________________ the set is said to be _______________.

- If the elements cannot be counted the set is said to be _________________.

- If a set has absolutely no elements, it is called the ______________ set or null set and is written in the form $\{\ \}$ or with the special symbol _____________.

- The notation ______________ is read "the set of all x such that …" and is called _________________-builder notation.

- The vertical bar that looks like __________________ is read "such that."

- Example: $\{x \mid x \text{ is an even integer}\}$ is read "the set of all x ______________ that x is an even integer."

- The ________________ (symbolized $\cup$, as in A $\cup$ B) of two (or more) sets is the set of all elements that belong to either ____________ set or the __________ set or to ___________ sets.

- The ________________ (symbolized $\cap$, as in A $\cap$ B) of two (or more) sets is the set of all elements that belong to _________________ sets.

- o The word ___________ is used to indicate union and the word ____________ is used to indicate intersection.

- o Suppose that a and b are two real numbers and that $a<b$. The set of all real numbers between a and b is called an __________________ of real numbers.

- o A ________________ interval such as $-3<x<6$ is read "x is greater than -3 and x is less than 6."

Complete the table below based on the information in Learn:

Type of Interval	Algebraic Notation	Interval Notation	Graph

- o In an open interval, _______________ endpoint is included.

- o In a closed interval, ______________ endpoints are included.

- o In a half-open interval, only ______________ endpoint is included.

- o The symbol for infinity, which looks like ________________ or ________________, is

  not a number. It is used to indicate that the interval is to include

  ________________ real numbers from some point on (either in the positive

  direction or the negative direction) without ____________.

- o When solving linear inequalities, ________________ or ________________ both sides

  of an inequality by a negative number causes the "sense" of the inequality to be

  ________________.

- o Rules for solving linear inequalities:

  1. Simplify each side of the inequality by removing any ______________ symbols and

      ______________ like terms.

  2. Use the ______________ property of equality to add the opposites of constants or

      variable expressions so that variable expressions are on one side of the inequality and

      constants are on the other.

  3. Use the ______________ property of equality to multiply both sides by the reciprocal of

      the coefficient of the variable (that is, divide both sides by the coefficient) so that the new

      coefficient is 1. If this coefficient is negative, reverse the sense of the inequality.

  4. A quick (and generally satisfactory) check is to select any one number in your solution

      and ______________ it into the original inequality.

- o Compound inequalities have ______________ parts and can arise when a variable or

  variable expression is to be between two numbers.

Normal Probability Distributions

Integrated Review

Area

o   Area is a measure of the _____________________ of (or surface enclosed by) a plane figure

and is measured in _____________________ units.

Units of Area

Basic U.S. Units of Area	Metric Units of Area
Square _____________________ (in.^{2})	Square _____________________ (mm^2)
Square _____________________ (ft^2)	Square _____________________ (cm^2)
Square _____________________ (yd^2)	Square _____________________ (km^2)
Square _____________________ (mi^2)	Square _____________________ (m^2)

o   Be sure to label your answers with the _____________________ units.

o   In triangles and other figures, we have used the letter _____ to represent the height of the

figure. The height is also called the _____________________ and is perpendicular to the

base.

o   Common area formulas:

_____________________ $A = \dfrac{bh}{2}$

_____________________ $A = l \cdot w$

_____________________ $A = s^2$

_____________________ $A = b \cdot h$

_____________________ $A = \dfrac{h(b+c)}{2}$

Solving Linear Equations: *ax* + *b* = *c*

o   The general procedure for solving linear equations is now a _________________ of the procedures stated in another section.

o   Steps for solving equations in the $ax+b=c$ format.

1.   _______________ like terms on both sides of the equation.

2.   Use the _______________ principle of equality and add the opposite of the constant b to both sides.

3.   Use the _______________ (or division) principle of equality to multiply both sides by the reciprocal of the coefficient of the variable (or divide both sides by the coefficient itself). The coefficient of the variable will become +1.

4.   Check your answer by _______________ it for the variable in the original equation.

o   Keep in mind that checking can be quite _____________________-_________________ and need not be done for every problem. This is particularly important on _____________. You should check only if you have time after the entire exam is _____________.

Example:

$$3x - 8 = 28 \qquad \text{Add 8 to each side.}$$

$$3x = 36 \qquad \text{Divide both sides by 3.}$$

$$x = 12$$

Working with Formulas

o   Formulas are general ______________ or principles stated ________________.

Below is a table of formulas and their meanings. You will use these in this lesson.

Formula	Meaning
$I=Prt$	The simple interest I earned by investing money is equal to the product of the ____________P times the rate of interest r times the time t in one year or part of a year.
$C=\dfrac{5}{9}(F-32)$	Temperature in degrees Celsius C equals $\dfrac{5}{9}$ times the ____________between the Fahrenheit temperature F and ____________.
$d=rt$	The ____________ traveled d equals the product of the rate of ____________ r and the time t.
$P=a+b+c$	The perimeter P of a triangle is equal to the sum of the ____________ of the three sides a, b, and c.
$F=kAv^2$	The lifting ____________ F is equal to the constant k by the area A and by the square of the plane's velocity v.

The Central Limit Theorem

Integrated Review

Ratios and Rates

o   Two meanings for fractions are:

  o   To indicate _____________________ of a whole.

  o   To indicate _________________.

o   A ratio is a _____________________ of two quantities by

  _____________________.

o   A ratio, 3 days to 4 nights, can be written in three ways:

  1.

  2.

  3.

o   Ratios have the following characteristics:

  o   Ratios can be _____________________.

  o   Whenever units of the numbers in a ratio are the same, the ratio has

  _____________________ units, otherwise known as an _____________________ number.

  o   When the numbers in a ratio have different units, then the numbers must be

  _____________________ to clarify what is being compared.

o   A _____________________ is a statement that two ratios are equal.

o   A proportion is _____________________ if the cross products are equal.

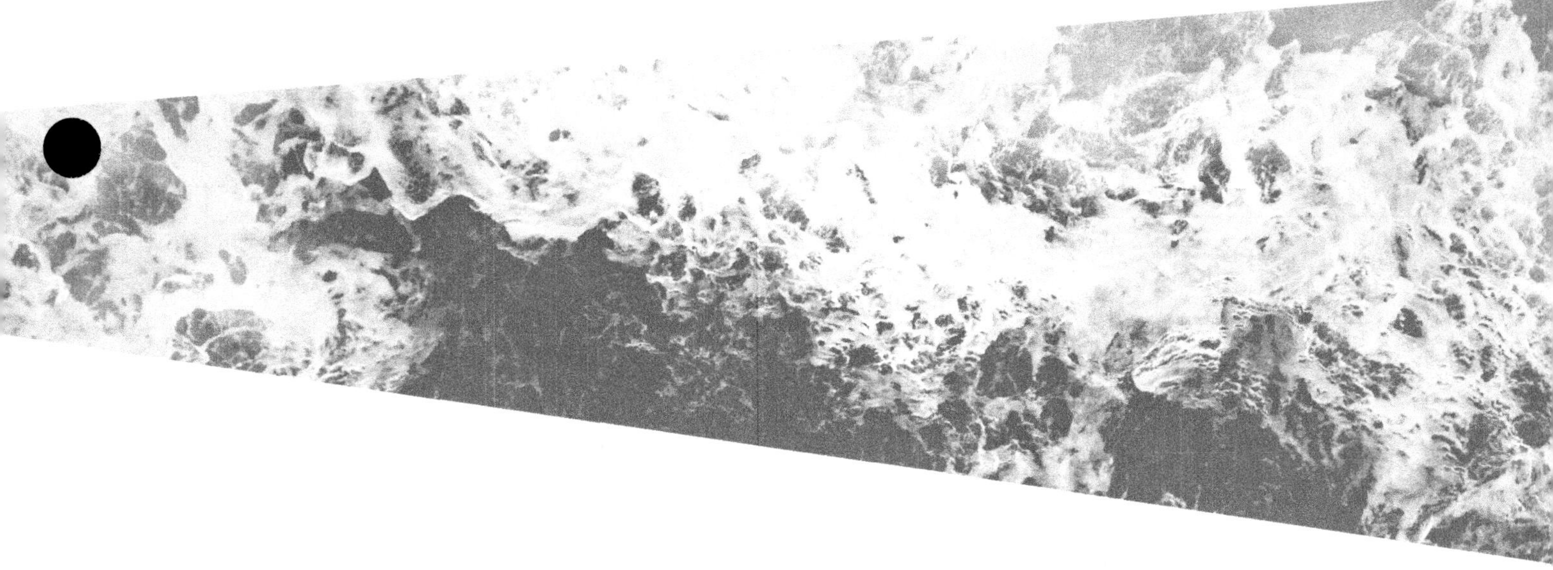

Hypothesis Testing
Integrated Review

Translating English Phrases and Algebraic Expressions

Addition	Subtraction	Multiplication	Division	Exponent (Powers)
Add	Subtract (from)	Multiply	Divide	Square of
Sum	Difference	Product	Quotient	Cube of
Plus	Minus	Times		
More than	Less than	Twice		
Increased by	Decreased by	Of (with fractions and percents)		
	less			

Key Words for Translating Phrases

- Remember that in each case "a number" or "the number" implies the use of a __________ (an unknown quantity).

- Be very careful when writing and/or interpreting expressions indicating ____________. Be sure that the subtraction is in the order indicated by the wording in the problem. The same is true with expressions involving______________.

- ____________________ and ____________________ are done with the values in the same order that they are given in the problem.

Order of Operations with Fractions and Mixed Numbers

o   To compare two fractions (to find which one is larger or smaller):

  o   Find the least common _____________________.

  o   Change each fraction to an _____________________ fraction with that

       _____________________.

  o   Compare the _____________________.

o   Rules for order of operations:

  o   Simplify within _____________________ symbols, like parentheses, brackets, and

       braces, starting with the _____________________ most group.

  o   Evaluate any numbers or expressions indicated by _____________________.

  o   Moving from _____________________ to _____________________, perform any

       multiplication or division in the order in which they appear.

  o   Moving from _____________________ to _____________________, perform any

       addition or subtraction in the order in which they appear.

Regression, Inference, and Model Building

Integrated Review

The Cartesian Coordinate System

- The Cartesian coordinate system (or the rectangular coordinate system) is based on a relationship between ______________ in a plane and ordered pairs of real numbers.

- For an equation written like $y = 5x + 7$, ordered pairs are written in the form of (__, _____).

- In an ordered pair, x is called the first ___________ and y is called the _________ coordinate.

- To find ordered pairs that satisfy an equation in two variables, we can choose any value for one variable and find the corresponding value for the other variable by ________________ into the equation.

- There is an _______________ number of such ordered pairs. Any real number could have been chosen for x and the corresponding value for y calculated.

- In an ordered pair of the form (x,y), the first coordinate x is called the ________________ variable and the second coordinate y is called the _______________ variable.

- Label the following information on the coordinate plane below: x-axis, y-axis, quadrants (4), and origin.

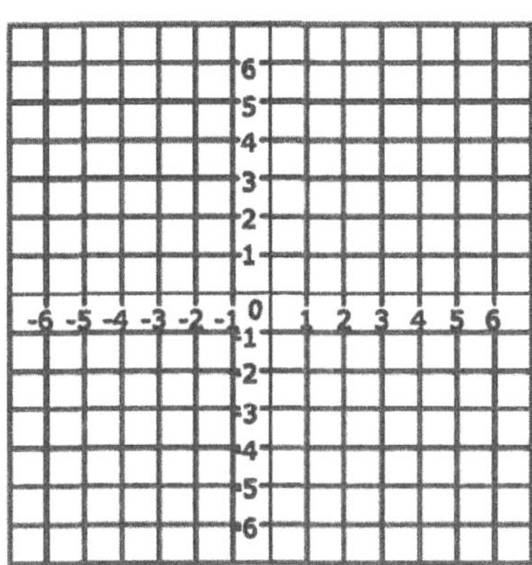

- In quadrant I, the x is _______ and the y is _____________________.

- In quadrant II, the x is ______________ and the y is ________________________.

- In quadrant III, the x is ____________ and the y is ________________________.

- In quadrant IV, the x is ____________ and the y is ________________________.

- There is a one-to-one ____________________ between points in a plane and ordered pairs of real numbers.

- Unless otherwise stated, assume that the grid lines are ____________ unit apart.

- To plot an x-coordinate, you will move____________________ for negative values and ____________________ for positive values.

- To plot any y-coordinate, you will move____________________ for negative values and up for ____________________values.

Graphing Linear Equations in Two Variables: $Ax + By = C$

- To find some solutions for an equation like $2 - 2x = y$, you will need to:

    1. choose arbitrary values for ________________

    2. find the corresponding values for ________________ by substituting into the equation.

- The standard form of an equation is ____________ where A, B, and C are real numbers and A and B are not both equal to ____________, is called the standard form of a linear equation.

- In the standard form ____________, A and B may be positive, negative, or 0, but A and B cannot both equal 0.

- Every line corresponds to some ________________ ________________, and the graph of every linear equation is a line.

- Two points determine a ________________.

- Steps to graph a linear equation in two variables:

    1. Locate any ________________ points that satisfy the equation. (Choose values for x and y that lead to ________________ solutions. Remember that there is an ________________ number of choices for either x or y. But, once a value for x or y is chosen, the corresponding value for the other variable is found by substituting into the equation.)

    2. Plot these ________________ points on a Cartesian coordinate system.

    3. Draw a ________________ through these two points. (Note: Every point on that line will satisfy the equation.)

4. To check: Locate a _______________ point that satisfies the equation and check to see that it does indeed lie on the line.

o   By letting $x =$_______________, you will locate the point on the graph where the line crosses (or intercepts) the y-axis. This point is called the y-_______________ and is of the form (___,y).

o   The x-intercept is the point found by letting $y =$__________ .This is the point where the line crosses (or intercepts) the x-axis and is of the form (x,_____).

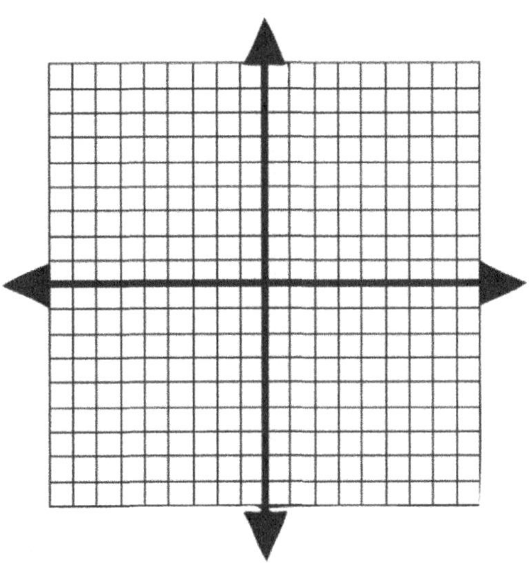

o   When the intercepts result in a point with fractional (or decimal) coordinates and _________________ is involved, then a _______________ point that satisfies the equation should be found to verify that the line is graphed correctly.

o   For real numbers a and b, the graph of $y = b$ is a _________________ line and $x = a$ is a _________________ line.

The Slope-Intercept Form: $y = mx + b$

o   For a line, the ratio of ______________ to ________________ is called the slope of the line.

o   Plot the rise and run between points on the line in the graph below:

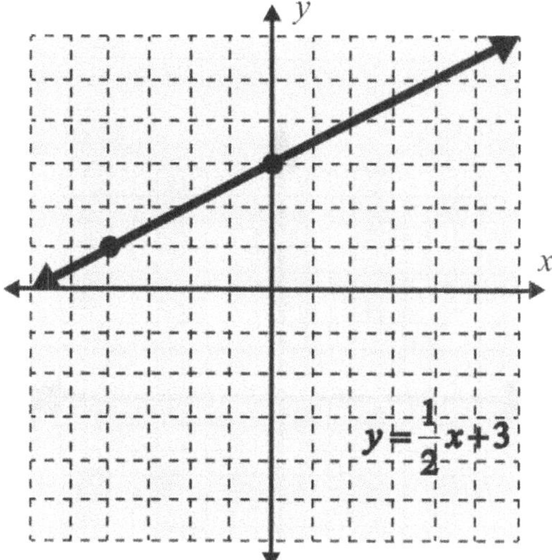

o   The concept of slope also relates to situations that involve ______________ of

______________.

o   ____________ Formula = $m = \dfrac{rise}{run} = \dfrac{y_2 - y_1}{x_2 - x_1}$

o   The variable that represents slope is ________.

o   The slope is the same even if the order of the points is __________________.

o   The coordinates must be __________________ in the same order in both the ______________

and the __________________.

o   Lines with positive slope go up (__________________) as we move along the line

from ______________ to ________________.

o Lines with negative slope go down(________________) as we move along the line

 from ______________ to ________________.

o If two points have the same ______________, such as (2, 8) and (7,8), then the line through

 these two points will be __________________.

o If two points have the same ______________, such as (2, 3) and (2,−2), then the line through

  these two points will be __________________.

o The following two general statements are true for horizontal and vertical lines:

 1. For horizontal lines (of the form $y = b$), the slope is __________________.

 2. For vertical lines (of the form $x = a$), the slope is __________________.

o *y=mx+b* is called the ______________-__________________ form for the equation of a line,

 where *m* is the __________________ and (0, *b*) is the __________________.

<u>Notes</u>

HAWKES
LEARNING
HAWKESLEARNING.COM

ISBN 1-944894-83-7
9 781944 894832